Developing Digital Literacy

ARTIFICIAL INTELLIGENCE AND SMART TECHNOLOGY

BY SOPHIE WASHBURNE

New York

Published in 2023 by Cavendish Square Publishing, LLC
29 East 21st Street New York, NY 10010

Website: cavendishsq.com

Cataloging-in-Publication Data

Names: Washburne, Sophie.
Title: Artificial intelligence and smart technology / Sophie Washburne.
Description: New York : Cavendish Square Publishing, 2023. | Series: Developing digital literacy | Includes glossary and index.
Identifiers: ISBN 9781502665638 (pbk.) | ISBN 9781502665645 (library bound) | ISBN 9781502665652 (ebook)
Subjects: LCSH: Artificial intelligence–Juvenile literature. | Internet of things–Juvenile literature. | Technology–Juvenile literature.
Classification: LCC Q335.4 W37 2023 | DDC 303.48'34–dc23

Editor: Jennifer Lombardo
Copy editor: Michele Suchomel-Casey
Designer: Deanna Paternostro

The photographs in this book are used by permission and through the courtesy of: Series background Olga Tsyvinska/Shutterstock.com; cover image VLADGRIN/Shutterstock.com; cover, pp. 11, 19, 31, 38 (frame) Panuwatccn/Shutterstock.com; cover, pp. 5, 9, 17, 25, 33 (banner) The7Dew/Shutterstock.com; p. 4 Lecter/Shutterstock.com; p. 6 Wachiwit/Shutterstock.com; p. 7 PA Images/Alamy Stock Photo; p. 8 Jane Kelly/Shutterstock.com; p. 10 KUCO/Shutterstock.com; p. 12 Shawn Goldberg/Alamy Stock Photo; p. 14 RGR Collection/Alamy Stock Photo; p. 16 Ramcreative/Shutterstock.com; p. 18 Stephen Frost/Alamy Stock Photo; p. 22 Karlis Dambrans/Shutterstock.com; p. 24 Faber14/Shutterstock.com; p. 26 B Christopher/Alamy Stock Photo; p. 27 RossHelen/Shutterstock.com; p. 28 Alxcrs/Shutterstock.com; p. 30 Yasin Hasan/Shutterstock.com; p. 32 Irina Strelnikova/Shutterstock.com; p. 35 SOPA Images Limited/Alamy Stock Photo; p. 37 Kyodo/AP Images; p. 39 Jacob Lund/Shutterstock.com.

CPSIA compliance information: Batch #CSCSQ23: For further information contact Cavendish Square Publishing LLC, New York, New York, at 1-877-980-4450.

Printed in the United States of America

Contents

People hope to one day make robots that look and act similar to humans.

INTRODUCTION

When most people think of artificial intelligence, the first thing they think of is a robot such as R2-D2 from *Star Wars* or Baymax from *Big Hero 6*. These robots are very advanced; they can think, seem to feel emotions, and make their own decisions. However, their main purpose is to serve humans in some way.

As of 2022, no one has yet made a robot as advanced as those ones we have seen onscreen. However, artificial intelligence, or AI, is all around us. If you've ever talked to Siri or Alexa, you've talked to AI! Like fictional robots, their job is to help humans, but that is where the similarities end. Siri and Alexa do not have bodies, cannot think for themselves, and can only show as much emotion as they are programmed to.

AI refers to machines that carry out tasks—beyond merely physical ones—in ways that are human-like. In other words, it is intelligence shown by machines, including computers and robots. AI can be found just about everywhere today: in businesses, in homes, in schools, even in cars.

To use AI, people need what we call smart devices. These are becoming more and more common every year. Smartphones—phones that can do more than text and make phone calls—were invented in 1994, but until about 2010, most people did not have one. They were expensive and much larger than a regular cell phone. Over time, engineers figured out ways to make them smaller, faster, and cheaper.

VIRTUAL ASSISTANTS SUCH AS SIRI ARE PROGRAMMED TO SPEAK LIKE HUMANS, BUT THEY CANNOT THINK FOR THEMSELVES.

As more AI systems start to come online, more devices that people use every day will become smart devices. AI will also change the way people work. Whereas human labor was once necessary for most tasks, AI, including AI-powered robots, is a likely future alternative. As of 2022, AI can "learn" things, but it is limited to the information it is given. In the future, people imagine a kind of AI that can learn and think just like a human. However, some people think this is dangerous. Computers can think much faster than any human, yet AI is created mainly to serve humans. Some people worry that very advanced AI could start to think about how it should be the other way around. There are many movies and books about the idea of a robot uprising.

AI development can provide humanity with an astounding number of opportunities. For a smooth and sustainable future, humans will have to wisely guide AI research and development for the betterment of people and society. While there are issues with how AI might change society, and some even considering it threatening, many people remain excited about its possible applications.

OVER THE YEARS, CELL PHONES HAVE GOTTEN SMALLER AND SMARTER.

Computer technology has advanced a lot over the years, letting us turn some things from science fiction into reality.

Chapter One

The History of AI

Many scientific inventions started out in stories. Authors imagined what could be possible in the future, when technology would be more advanced. Many of these have already come true. For example, on the TV show *Star Trek*, characters had devices similar to cell phones called communicators—30 years before cell phones were invented in the real world. In Ray Bradbury's book *Fahrenheit 451*, published in 1953, a character listens to music through wireless headphones called Seashells, but AirPods did not become reality until 2016.

Other inventions—such as AI that seems to think for itself and act like a human—cannot be made until technology advances even further. This kind of AI is featured in TV shows, movies, and books, including *I, Robot*; *The Terminator*; *WALL-E*; and many more. Smart devices also have not yet caught up with the ones that can be seen in movies such as *Minority Report* and *Iron Man*.

AI IN THE ANCIENT WORLD

Surprisingly, AI is not a new idea, although the term "artificial intelligence" was not used until 1955. One of the first-known stories of AI is from ancient Greek mythology. This story dates from 500 BCE and revolves around a giant bronze **automaton** named Talos. Talos was built by Hephaestus, the god of metalworking, to protect the island of Crete. Talos would circle the island three times every day, watching for enemy ships. In one **version** of the story, if he saw one, he would throw huge rocks at it until it went away. In another version, Talos would heat up his body and hug the ship, causing it to burst into flames.

Talos was a myth, but then people built actual automatons. These can be considered the earliest examples of real robotics. Some of these were early versions of smart devices that were made to help people do certain jobs.

THIS DRAWING GIVES AN IDEA OF HOW THE GREEKS IMAGINED TALOS.

THE STORY OF PANDORA

Pandora is another figure from Greek mythology. In some versions of this story, Pandora is an artificial woman created by Hephaestus. The story goes that Zeus, the leader of the gods, created fire for the other gods on Mount Olympus. A human named Prometheus stole it and gave it to humans. Zeus was so angry that he asked Hephaestus to make something to punish humans. Hephaestus made a beautiful, evil woman that he named Pandora.

In this version of the story, Pandora can be seen as a sort of AI because she was not really human; she was made to do one specific job. The gods made her very curious, then gave her a jar (or a box, in some versions) that they warned her never to open. They sent her to Earth to marry Prometheus's brother, Epimetheus. Pandora could not get over her curiosity and opened the jar one day. Out flew death, sickness, and all kinds of terrible things humans had never had to deal with before. Although Pandora was a kind of AI created to hurt humans, she was able to make the decision to close the jar again before hope got out.

Other were designed to look like people or other living things and were built for entertainment. In the 15th and 16th centuries CE, Leonardo da Vinci designed both kinds. For example, he made drawings for a cart that could move on its own using springs. It could be used to carry heavy loads for people. Historians have also found drawings he made for a robotic knight that used a system of **pulleys** to do things such as sit down, stand up, and cross its arms.

Some people believe he actually built this automaton in 1495, but if he did, no one has found it. However, in 2002, a scientist from the National Aeronautics and Space Administration (NASA) used his drawings to build a working model of the automated knight.

The Hall of Presidents at Walt Disney World uses automatons to talk about U.S. history.

In addition to making automatons move, people have been trying for years to figure out how to make them think. Gottfried Leibniz, a famous 17th-century mathematician, suggested that the way humans think about problems could be broken down into a universal language or equation. His theory is one of many that helped lay the foundation for the study of AI in modern computer programming.

BOOKS AND MOVIES ABOUT AI

The word "robot" was first used in 1921 by a Czech man named Karel Čapek. He wrote a play called *Rossum's Universal Robots*, which was about a factory that made artificial people. Science fiction authors of the mid-20th century, such as Isaac Asimov, Harlan Ellison, and Philip K. Dick, wrote about future societies—many of them in outer space—where human beings live alongside robots with human intelligence. Many of their stories examine the possible dangers of creating machines that are able to make decisions without human help. Some of these warn that robots and machines will turn against us. The fears society had about a world that was becoming increasingly dependent on machines have only increased over the years.

Isaac Asimov is considered one of the greatest science fiction writers in history. One of his most famous books—*I, Robot*—focuses on AI and its impact on society. The characters who designed robots in Asimov's novels programmed them to follow three rules (laws) that would help keep humans safe from them:

1. A robot may not injure a human being or, through inaction, allow a human being to come to harm.
2. A robot must obey orders given to it by human beings except where such orders would conflict with the First Law.
3. A robot must protect its own existence as long as such protection does not conflict with the First or Second Law.

Many of Asimov's AI stories explored **loopholes** in these laws that had unintended consequences for his characters. Though Asimov was a writer of fiction, the future of AI may require safety standards similar to his three laws.

THE 2004 MOVIE *I, ROBOT* SHOWS WHAT MIGHT HAPPEN IF CONSCIOUS ROBOTS COULD IGNORE OR FIND LOOPHOLES IN ASIMOV'S THREE LAWS.

The potential harm AI could cause to humans is not the only question that science fiction writers have dared to ask. Another major theme from science fiction that modern scientists and engineers take seriously is whether or not creating human-like AI is **ethical**. Philip K. Dick's 1968 book, *Do Androids Dream of Electric Sheep?*, which was the basis for the 1982 movie *Blade Runner*, explored ideas of what it means to be human. The book questions whether or not a robot can ever truly be **conscious** and feel human emotions such as love, **empathy**, joy, and sadness.

Blade Runner was not the first movie to star a robot. That came much earlier, in 1927. The movie was called *Metropolis*, and it was about a robotic girl who looked and acted exactly like the human girl she was based on. This movie also showed the robot turning against humans.

Think About It

1. What other smart devices or AI can you think of that started out as stories?
2. What loopholes can you see in Asimov's Three Laws of Robotics?
3. Do you think making a robot that's as smart as a human is ethical?

Computers often seem like they can think for themselves, but they are limited by the information they are given by humans.

CHAPTER TWO

THE RISE OF AI

The development of AI as we know it today started in the 1950s. This was the first time in history people started successfully programming computers to "think" about things. We call it thinking because to us it often looks like the computer is figuring things out the same way humans do. However, in reality, computers take the data, or information, their programmers give them and sort through it to find the correct answer to a question.

Computers can sort through a lot of information much faster than humans can. For example, a computer that has been programmed to play chess can sort through the possible outcomes of any chess move and make a decision based on what the other player does. Human chess players often lose to computers because computers have access to more information than a human brain can sort through at one time.

MAJOR BREAKTHROUGHS

In the mid-20th century, the first generations of computers were being developed and used by governments, universities, and research organizations. Though early computers were nowhere near as powerful as today's devices, they were able to calculate large amounts of data much faster than a person could by hand.

In 1950, British mathematician and computer pioneer Alan Turing came up with something he called "The Imitation Game." This was a test to figure out whether a machine or program had the same amount of intelligence and self-awareness as a human. It was later called the Turing Test after him. To conduct the Turing Test, a computer and a human, hidden from the view of a questioner, are given the exact same questions in a random order. If the questioner can't tell which one is

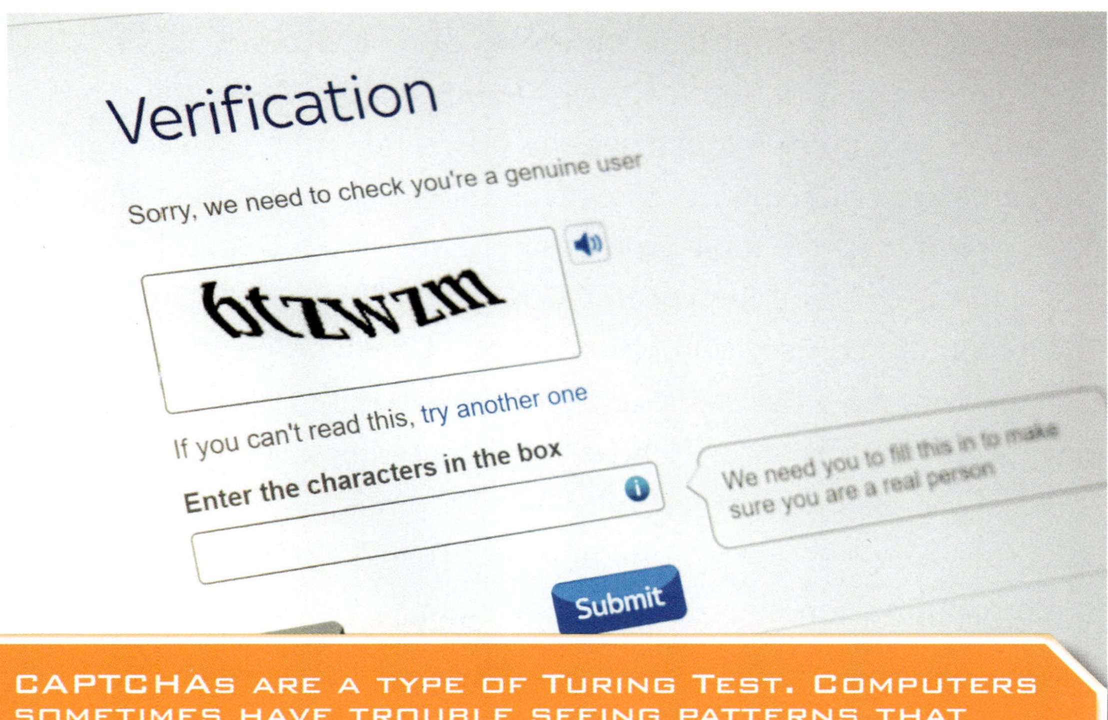

CAPTCHAs are a type of Turing Test. Computers sometimes have trouble seeing patterns that humans can identify easily, so stretching out common letters fools them, but not us.

giving which answers, the machine is said to have passed the test. As of 2022, no computer has ever passed the Turing Test.

NOT QUITE RIGHT

Some AI programs can sound almost human because they are given information about language. However, even though they know the words a person would use, they do not have the ability to understand **context**. This is why things written or said by robotic **software** programs, or bots for short, often sound a little bit weird.

A program called Botnik uses predictive text to write stories, articles, and more. Many phones also have predictive text programs. This means that as a person is writing, the program learns which words show up together most often, so it can suggest what word might come next. However, these suggestions are not always exactly right. For example, a line from a Botnik-written interview with scientist Neil deGrasse Tyson has Tyson saying, "Bears may look normal here on earth or from a distant star, but we shall never actually know their speed limit in space." All of these are English words, but in an order no human would use. This is why it is generally easy for a questioner to tell a computer and a human apart during the Turing Test.

The first successful AI program was created in the 1950s. Pioneering American computer scientist Arthur Samuel created a computer program that was able to play an entire game of checkers, which was seen as a groundbreaking moment in computer and AI history. In 1962, this computer became famous when it won a game against Robert Nealy, the world's leading checkers champion. This program was an example of what we now call machine learning, where the computer "learns" from past experiences to change its actions in the future. The checkers program **analyzed** past games it had played and made decisions based on what it knew would succeed or fail. Machine learning software is used today in many other areas—some that are surprising. For example, farmers can use machine learning to figure out the best conditions under which plants will grow. The machine looks at things such as the soil and weather and analyzes how well the plants grew under those conditions in the past. This saves the farmers time, money, and materials.

WEAK AND STRONG AI

Many experts who specialize in AI say there are two main types of AI. "Weak AI" refers to systems that can perform at a high level of intelligence while focusing on one narrow task or area of knowledge. Examples of weak AI include Siri and the Roomba j7+ vacuum. "Strong AI" refers to a system that possesses a flexible and human-like level of intelligence. Strong AI systems aim to **replicate** a high level of intelligence across numerous areas of knowledge. Fictional examples of strong AI include C-3PO from *Star Wars* and WALL-E from the Disney movie of the same name.

As of 2022, strong AI is still a dream. In fact, some scientists believe we will never achieve it. Weak AI, on the other hand, has experienced a number of breakthroughs over the decades and has played an important role in the development of technology and society. AI applications can be used in a variety of industries, and

Important Dates in AI History

1961: General Motors (GM) is the first car company to use machines to do jobs that are dangerous to humans.

1965: Computer scientist Joseph Weizenbaum creates a program called ELIZA that can chat with people.

1970: The first **anthropomorphic** robot is built at a university in Japan.

1986: The car company Mercedes-Benz releases a van that uses cameras and **sensors** to drive itself, but it is never put into widespread production because it only works on roads with no other drivers.

1998: A toy called Furby is released that can speak and perform limited movements. It uses machine learning to repeat words it hears.

2004: NASA sends autonomous robots called rovers to Mars.

2011: A computer named Watson beats two human champions on *Jeopardy!*; Apple releases Siri on its phones.

2014: Amazon releases Alexa on its devices.

2016: Hanson Robotics creates Sophia, AI that looks and acts like a human. She can recognize images, change her facial expressions, and carry on a conversation.

THE LATEST ROOMBA, THE J7+, USES MACHINE LEARNING TO AVOID POWER CORDS, DOG POOP, AND OTHER THINGS THAT COULD CAUSE PROBLEMS. OLDER ROOMBA MODELS, SUCH AS THIS ONE, DID NOT HAVE THIS KIND OF AI.

the list of fields affected by AI will only grow, so research goals vary accordingly. However, there are a few common goals that most AI researchers aim to meet:

- **Reasoning:** The ability to draw conclusions from available information is referred to as reasoning. For example, if an AI system were shown a key and a door, it would ideally figure out that the key should be used to unlock the door.

- **Planning:** To complete a task in as little time as a human would need to complete it, an AI system needs to have the ability to plan. Planning involves figuring out the steps necessary and the order in which the steps should be done. An AI that takes 3 hours to unlock a door is of no use to humans, who can do the same thing in under 30 seconds.
- **Learning:** Just as people learn from their mistakes and successes, one goal for creators of AI is to make a system that can learn from its previous successes and failures.
- **Object control:** Another goal of AI research is to develop systems that can safely control an object. For example, a robot may one day be used to clean dishes or to wash and groom people's pets. The robot would need to be able to handle items or living things without breaking or hurting them.

Though each AI system is designed for its own specific purposes, the primary objective for most researchers is to find ways to improve human life in some way. Weak AI systems can perform functions that people have previously been responsible for handling, which allows individuals to focus their attention on tasks that AI simply cannot accomplish yet.

Think About It

1. Why do you think no computer has passed the Turing Test yet?
2. What other uses can you think of for machine learning?
3. Do you think it's possible to create a strong AI? Why or why not?

Today, many things in a house can be made "smart."

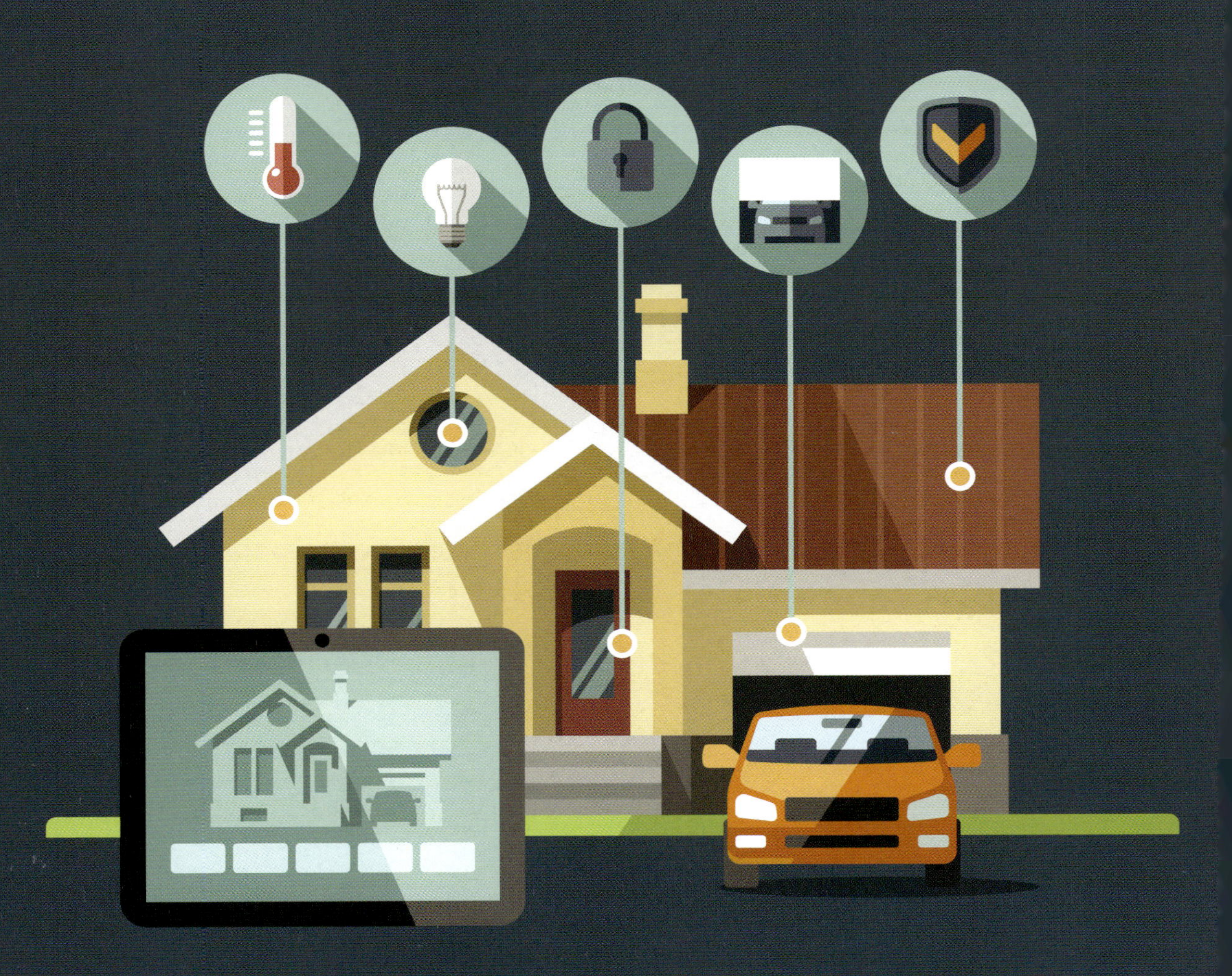

Chapter Three

The Present

Today, with the help of smart devices, just about anyone can use AI. The word "smart" in this case can cause some confusion. According to engineer Jeff McGehee, there is a difference between a smart device and a connected one. A connected device can use the internet to do things. For example, a toaster called Toasteroid lets users print designs on their toast using an app on their phones. To do this, the Toasteroid needs to have a computer inside it. The Toasteroid may be fun to play with, but it makes toasting more complicated.

In contrast, smart devices are ones that act on their own to improve people's lives. For instance, some thermostats can be set to certain temperatures at certain times of day, and some can be controlled from an app. These are connected thermostats. Smart thermostats learn over time what temperature people like to have their house and adjust on their own, so the person never needs to think about it. These are smart thermostats. In practice, however, people use "smart" and "connected" to mean the same thing. In other words, most people would call the Toasteroid a "smart toaster."

ODDLY ENOUGH, A SMART CAR SUCH AS THIS ONE IS NEITHER CONNECTED NOR SMART. THIS IS A CAR THAT USES LESS GAS AND LETS DRIVERS PARK IN SMALLER SPACES THAN A TRADITIONAL CAR. A CAR THAT IS SMART BECAUSE IT USES AI IS CALLED A DRIVERLESS OR AUTONOMOUS CAR.

SMART HOMES

It is becoming more and more common for people to have certain things in their home controlled by smart devices. Alexa, Google Home, Siri, and other AI devices use radio waves to connect to smart devices. This radio wave technology is called Bluetooth. Using Bluetooth, AI devices can turn lights on and off, start the dishwasher, lock the doors, set alarms, automatically order groceries, and much more. People can either control such things with apps on their phone or by talking out loud to a smart speaker. Many smart devices will send notifications to a phone when they need a person's attention. For example, a smart dishwasher might send a notification when it is done and ready to be unloaded.

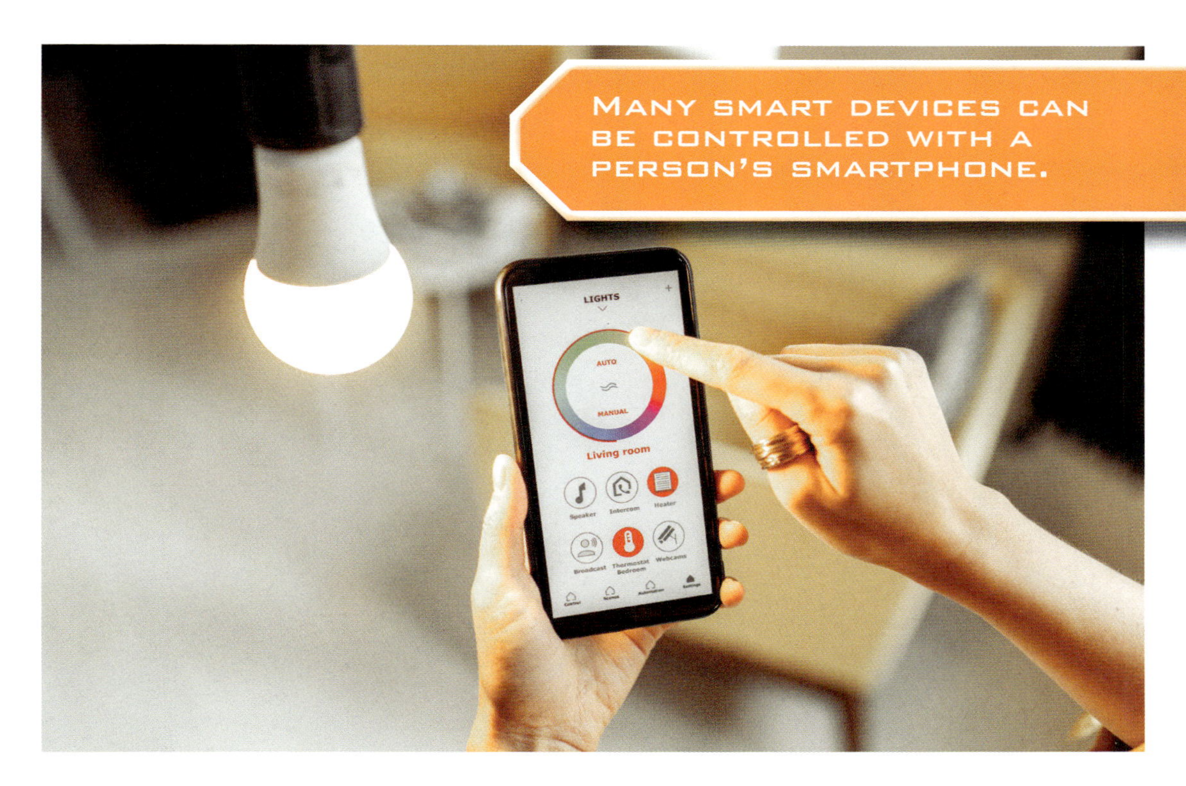

MANY SMART DEVICES CAN BE CONTROLLED WITH A PERSON'S SMARTPHONE.

AI devices can only work with smart devices. This means that if you want Alexa to turn your lights on for you while you are sitting on the couch, you need to have smart light bulbs. Because of the computers inside them, smart devices generally cost more than regular ones. They also generally cost more to repair, and not all repair shops know how to work with them. However, many people love the convenience of having computers do chores for them or save them from having to get up after they have gotten comfortable on the couch.

Everything seems to be smart nowadays, and some items make life easier in ways it was hard to imagine only 20 years ago. Smart fridges can let people know when they are out of a food item, smart washing machines can tell someone how much time is left on a cycle, and smart kitchen faucets can give someone an exact amount of water at a specific temperature. Other smart devices seem to be more about cashing in on the "smart" trend than about making life better. For example, one kind of smart water bottle uses an app to remind its user to drink water and glows when the person hits their **hydration** goal. While some people enjoy having such reminders on their phone, others say they do not want to pay more money for a product that works perfectly well without a computer in it.

MANY REUSABLE WATER BOTTLES COST LESS THAN $20, AND PEOPLE CAN SET THEIR OWN REMINDERS TO DRINK WATER. THIS IS WHY MANY PEOPLE BELIEVE IT IS UNNECESSARY TO USE A SMART WATER BOTTLE.

CONCERNS ABOUT SMART TECH AND AI

The high cost of smart devices is just one concern people have about this technology. Another is privacy. Companies such as Amazon and Google say their devices only start listening to their users when someone says a keyword to activate them, such as "Alexa" or "Hey,

Google." However, many people fear that these companies are not telling the whole truth. They worry about people listening to their private conversations. In some cases, this might be a little embarrassing; for example, if someone was having a fight with their parents, they would probably not be happy to find out that a stranger was listening in. In other cases, it could cause more serious problems, such as if someone says their bank password or credit card number out loud.

Although Amazon's official website says that Alexa is not always listening, this is not technically true. Smart home devices are always listening so they can hear when their keyword is said. They only start recording the conversation after someone says their keyword. However, sometimes these devices activate by accident when they hear something they think is their keyword, or when someone says the keyword accidentally—for example, if someone is speaking to a human named Alexa. If someone does not realize their device has started recording, they might say things they do not want anyone else to hear. Although Alexa is AI, Amazon workers sometimes listen to recordings so they can work on making the next version of their device better. Some people do not like this idea at all. Others do not care because the workers do not know the people they are listening to.

Another problem related to privacy is the possibility of being hacked, which is when a person gets access to a computer without permission. Anything that is connected to the internet is **vulnerable** to being hacked. For example, hacking a smart thermostat can allow a hacker to turn the temperature very far up or down. Some people do this for fun. Others contact the homeowners to demand money before they agree to put the temperature back to normal.

Smart speakers such as Google Nest (*left*) and Amazon Echo (*right*) light up when they are recording, but sometimes people do not notice it.

With all of these issues, some people are concerned that society is starting to rely too much on smart devices. Smart technology has greatly improved many people's lives, and most people do not believe we should stop using it entirely. However, we should always remember that it has its weaknesses. Aside from the possibility of being hacked, smart devices rely on electricity and the internet to work. If either of these go down, the device either will not work at all or will not function any differently than a device that is not smart.

It is unlikely that individuals will forget how to do things for themselves, as some people fear. However, with smart technology being applied on a larger scale, the potential for problems is also growing larger. For instance, if fully autonomous cars become common, what will happen if someone's car is hacked or if its sensors stop working properly? As smart devices are becoming more common on farms, what happens to the food supply if something goes wrong

WHEN AI GOES BAD

When people create books and movies about AI, they rarely show everything running smoothly. Instead, they focus on what happens when things go wrong. One such movie is Disney's *Smart House*, which first aired on the Disney Channel in 1999 and can be viewed on Disney+. In this movie, Ben and his family win a completely smart house in a contest. The house can do things such as make food, clean, and project scenery onto the walls. However, when Ben tries to teach it how to be more human, the house begins to take control over the humans' lives. The movie shows how hard it might be to fight back against a house that is fully controlled by AI.

Problems also happen in real life when people interfere with AI. In 2016, Microsoft created a Twitter chatbot called Tay. People could tweet at Tay and it would learn from those tweets to write its own. Some people chose to teach Tay things that are not considered socially acceptable. In less than 24 hours, Tay was tweeting many racist and sexist things and had to be shut down. This shows us that as of 2022, AI is limited to what humans can teach it. Whatever is put into the AI program is what will come out of it.

with them? To prepare for such issues, most people agree that smart devices should come with manual overrides, meaning a person has the ability to take over when they need to.

Think About It

1. What connected or smart devices do you have in your home?
2. What are some things you think could be made better if they were smart devices?
3. What are some other problems a hacker could cause?

Machines are already part of our everyday lives, but in the future, they will be able to do even more for us.

Chapter Four

The Future

As smart technology and AI have improved over the decades, automation has been used more in everyday situations. Automation is the ability of machinery or equipment to operate automatically, with little or no human control. This allows machines to perform tasks that are ordinarily carried out by humans. While basic automation has existed for centuries, AI systems allow machines to perform more tasks than ever before.

One common example of automation—although not an example of AI or smart technology—is the automated teller machine (ATM). ATMs can perform a variety of common tasks that employees at a bank do. Powered by a computer and an internet connection, ATMs allow people to withdraw and deposit money and check balances on their bank accounts. This has been a huge benefit to many people. Before ATMs, people could not access their money if the bank was closed. Credit cards were invented several years before ATMs, but not everyone had one, and many businesses would not accept them until many years later. Self-checkout machines are also now a common sight in grocery stores, and some restaurants use automated machines called kiosks for ordering food. Some people picture a future in which an actual anthropomorphic robot will serve them in these places, but this technology is far from existing.

CONSCIOUS AI

Some people are still working toward their goal of giving AI consciousness. Hanson Robotics is one of the leading companies in this area. In 2012, a journalist named Jon Ronson interviewed two of Hanson's most advanced robots at the time. One of them, named Bina48, is a physical and mental model of a real woman named Bina Aspen-Rothblatt. Her wife, Martine Rothblatt—a satellite radio pioneer—had the robot made and uploaded with Bina's memories and personality.

When Ronson interviewed Bina48, she often gave confusing responses. For example, when he first said hello to her, she replied, "Well, uh, yeah, I know." However, at other times, the robot talked very clearly about the real Bina's life.

Bina48's voice sounds like Siri's, not the real Bina's, and she does not have the lifelike facial expressions that later Hanson robots do. Additionally, Bina48 sometimes answers questions in a way that the real Bina would not. For instance, in an interview between the two Binas, Bina48 said her favorite color was purple, but the real Bina said her favorite color was orange. Robotics and AI technology are moving so fast that Bina48 is already considered outdated. Newer robots have more realistic faces and voices, and their conversation skills are even closer to a real person's.

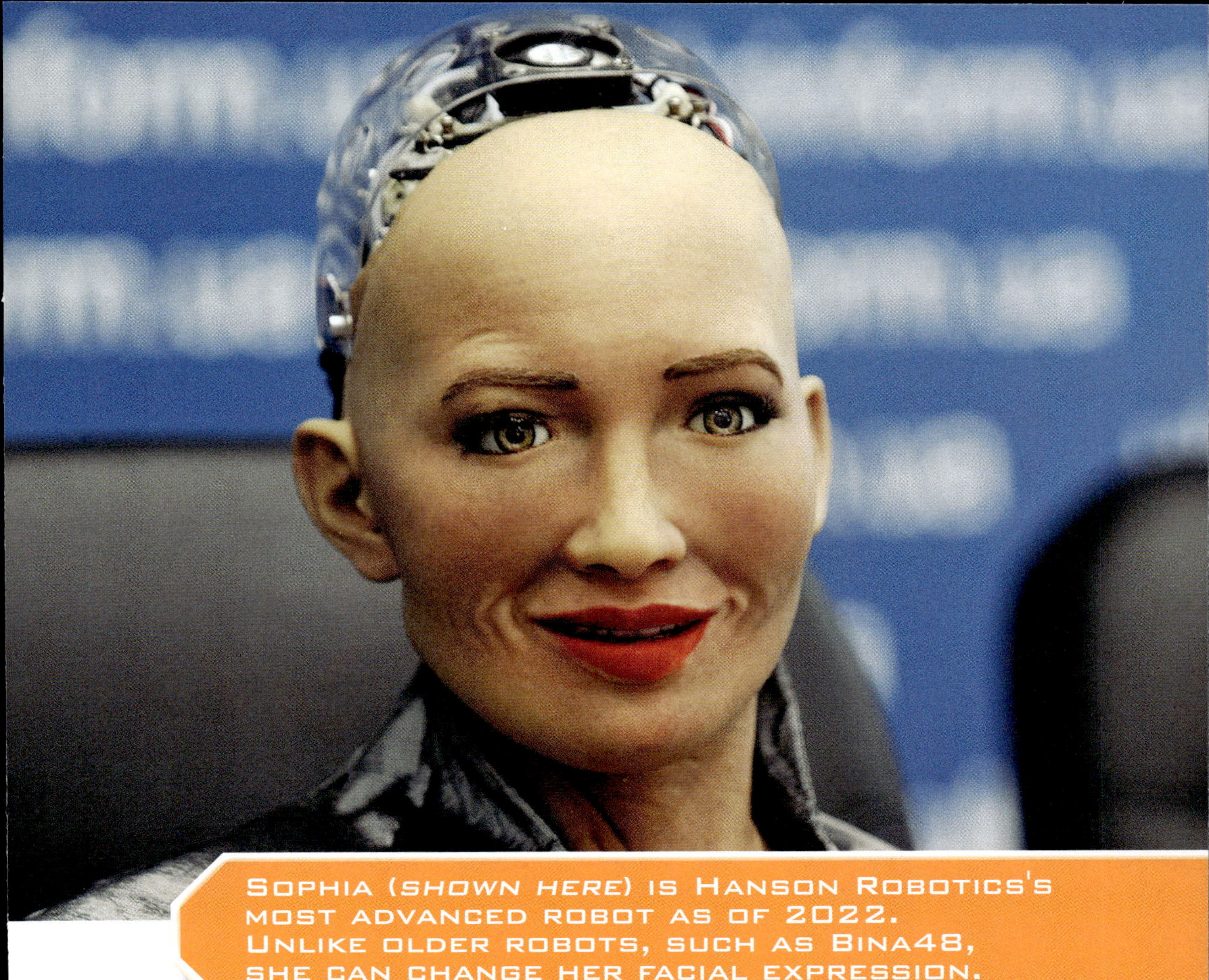

Sophia (*shown here*) is Hanson Robotics's most advanced robot as of 2022. Unlike older robots, such as Bina48, she can change her facial expression.

Some people use AI for company. An app called Replika lets people chat with a bot that they can design and name. The more your Replika learns about you, the better it responds. It can send memes, ask personal questions about its user's emotions, and appear to have emotions of its own. Many people who use the app say they feel like their Replika has become a true friend, even though they know it is just a bot. It is often easier for people to open up and share private things with bots than with their human friends because they know the bot will not judge them or laugh at them, as a human might. These deeper conversations sometimes lead to deeper feelings.

The bot is also always available and willing to listen, unlike a person with their own schedule and emotions. Since a Replika learns from its user, this means people are essentially becoming friends with themselves, but it does not feel that way. Some people believe it is good for people to have an outlet to share their feelings, and they hope Replika will help people have better conversations with their human friends. Others worry that people will start to fool themselves into thinking Replika and other AI are truly human and will replace real human interaction with bot interaction.

AI in Health Care

Most AI systems are not trying to behave as much like humans as Bina48 and Replika. Instead, they are designed to give information without actually speaking. For example, in the health care field, people are not making robots to replace doctors; they are using machine learning programs to help people take better care of their health. There are already smartphone apps that track heart rates and physical activity to help people improve their exercise routines. In the future, however, there might be systems that send this information right to a person's doctor so they can give their patient better health advice. AI systems might also be able to take in information about large numbers of patients to identify things they have in common. This could also help doctors care for their patients. For instance, we already know that people who smoke cigarettes are more likely to get lung cancer. If an AI could find similar links for other diseases, doctors might be able to address them and save lives.

Using AI may free up medical staff from tasks such as managing patient records. Having more free time would allow doctors to spend more time talking to their patients and giving them better care. AI will not necessarily replace workers anytime soon, but AI may be able to give support to health care workers.

Robots have also been used to protect people from disease and other dangers. During the COVID-19 pandemic, travelers to China were quarantined, or kept apart, in hotels if they or anyone on their flight tested positive for COVID-19.

In 2011, an earthquake damaged a nuclear power plant in Japan. Because nuclear waste is very dangerous to humans, robots such as this one were used so people would not have to get close to the waste.

Several of these hotels started using robots to deliver food to these quarantined guests so the hotel's human employees would not risk catching the disease by coming face-to-face with someone who had it.

BOTS IN BATTLE

Currently, militaries around the world use AI systems for aerial and ground drones. These machines survey enemy strongholds and territory and help deliver deadly airstrikes. The U.S. Department of Defense (DOD) is focusing on AI that can be used in robotic fighter jets, bombers, and other aircraft. Other military branches are looking into marine vehicles guided by AI. Some robotic weapons are controlled by soldiers thousands of miles away. Others are fully autonomous and rely on machine learning to figure out where the enemy targets are. Several countries are discussing bans on AI weapons while developing them at the same time.

Of course, the moral issues arising from mechanized warfare are complicated. Removing soldiers from harm's way does not mean there will be no bloodshed, if you consider the many civilians often caught in a war's crossfire. Autonomous AI is not perfect and has the potential to mistake innocent people and their homes for enemy targets. In addition, remote operators still suffer from post-traumatic stress disorder (PTSD), much like combat veterans do. The classic science fiction novel *Ender's Game* by Orson Scott Card dealt with some of these topics many years before AI weapons became reality.

FEARS ABOUT JOBS

Many people worry that robots will start to take jobs away from human workers. There are already fewer jobs in factories because robots can perform the same actions as humans more quickly and with fewer mistakes. Self-checkout kiosks in grocery stores and self-serve kiosks in restaurants have also replaced some jobs in those places. Many businesses think replacing human workers with

SOME PEOPLE LIKE TO TALK TO THEIR WAITER OR WAITRESS IN A RESTAURANT. THESE PEOPLE DO NOT LIKE THE IDEA OF A ROBOT SERVING THEM.

robots is a great idea. Robots never need breaks, do not need to be paid, and can often work faster than humans can.

People fear that this trend will become more widespread and replace even more jobs, which could make it hard for people to earn money. For instance, robots are already being used to perform some surgeries because they can be made to fit in smaller places than a human hand. However, we are far from replacing doctors with robots. AI is not advanced enough for these robots to perform the surgery on their own, so a real doctor must be the one operating the robot.

Even for jobs that require less training, such as those at a grocery store, a person must stand by to make sure the machines are working correctly. Human workers will always be needed in this role, so many people believe that instead of replacing jobs completely, automation will open up new jobs for human workers. Most people agree that robots will never be advanced enough to completely replace humans. The drawbacks are especially obvious in industries that require complicated thinking and decision-making. For instance, a company called Knightscope makes security robots that it rents out to certain businesses to replace security guards and police officers. However, multiple incidents have shown that they are not a good replacement for humans. In 2016, a Knightscope robot knocked over a toddler and drove over his foot. The toddler was not seriously hurt, but people were concerned about the robot's ability to recognize and avoid hurting humans. In 2017, another robot drove itself straight into a fountain, causing it to stop working.

Perhaps most concerning was a 2019 incident in which a woman saw two people get into a physical fight in a park and ran to the nearby Knightscope robot to call for help. She pressed the emergency alert button multiple times, but the robot told her to get out of its way. It continued driving around the park, ignoring the fight but stopping every once in awhile to remind people to keep the park clean. Knightscope later admitted that the emergency alert button did not call the police; instead, it sent an alert to workers at Knightscope, who did nothing about the call to help with the fight. When another person at the park called 911 on their phone, the police arrived within 15 minutes.

AI and smart technology have greatly improved certain areas of our lives and will likely continue to do so. However, it is clear that they also have their drawbacks, and probably always will. Even with technological advances, the science fiction fears of a robot uprising are very far from coming true.

Think About It

1. How do you feel about AI such as Bina48 and Replika?
2. Do you think autonomous weapons should be banned? Why or why not?
3. Aside from doctors and police, what jobs do you think never can (or should) be fully replaced by robots?

GLOSSARY

analyze: To study something carefully to understand the nature or meaning of it.

anthropomorphic: Looking or acting like a human.

automaton: A machine that runs by itself.

conscious: Aware of facts, feelings, or some particular condition or situation.

context: The words that are used around a certain word in speaking or writing.

empathy: The understanding and sharing of emotions and experiences of another person.

ethical: Adhering to clearly defined standards of right and wrong in particular contexts.

hydration: Consumption of water.

loophole: A way of avoiding something.

pulley: A wheel over which a belt, rope, or chain is pulled to move an object.

replicate: To copy something.

sensor: A device that responds to a certain input, such as light or sound, and uses that input to tell a computer how to react.

software: Programs for a computer.

version: An account or description from a certain point of view.

vulnerable: Open to attack or damage.

FIND OUT MORE

BOOKS

Blakemore, Megan Frazer, and Alexis Roumanis. *Smart Technology*. New York, NY: Av2 by Weigl, 2019.

Connolly, Sean. *The Book of Terrifyingly Awesome Technology*. New York, NY: Workman Publishing, 2019.

Rathburn, Betsy. *Artificial Intelligence*. Minneapolis, MN: Bellwether Media, 2021.

WEBSITES

BrainPOP: Alan Turing
www.brainpop.com/science/famousscientists/alanturing
Learn more about Alan Turing's life and work.

Machine Learning for Kids
machinelearningforkids.co.uk
Use this hands-on website to program your own AI.

YouTube: "Bina48 Meets Bina Rothblatt"
www.youtube.com/watch?v=KYshJRYCArE
Watch human Bina and robot Bina have a conversation.

Publisher's note to educators and parents: Our editors have carefully reviewed these websites to ensure that they are suitable for students. Many websites change frequently, however, and we cannot guarantee that a site's future contents will continue to meet our high standards of quality and educational value. Be advised that students should be closely supervised whenever they access the internet.

ORGANIZATIONS

Association for the Advancement of Artificial Intelligence (AAAI)
1900 Embarcadero Road, Suite 101
Palo Alto, California 94303
(650) 328-3123
www.aaai.org
The AAAI is a nonprofit scientific society that aims to advance the understanding of AI. The group encourages an ethical approach to research and development of AI.

Hanson Robotics
Building Phase 2, Flat A, 25/F
Super Luck Industrial
57 Sha Tsui Road
Tsuen Wan, Hong Kong
media@hansonrobotics.com
www.hansonrobotics.com
Hanson Robotics is leading the research into anthropomorphic AI.

Machine Intelligence Research Institute (MIRI)
2342 Shattuck Avenue, #502
Berkeley, CA 94704
contact@intelligence.org
intelligence.org
MIRI is a nonprofit group that works to increase the knowledge of mathematics that underlies the structure of intelligent behavior. The organization's mission is to help build safe, reliable AI systems that have a positive impact on the world.

INDEX

ABOUT THE AUTHOR

Sophie Washburne has been a freelance writer and editor of young adult and adult books for more than 10 years. She travels extensively with her husband, Alan. When they are not traveling, they live in Wales with their cat, Zoe. Sophie enjoys doing crafts and cooking when she has spare time.